Higgs Field Unveiled

God's Field at Creation

Gerald W. Scanlon

Contents

Preface

For millennia, there has been a quest to find a unified theory that would explain the laws that govern the universe. What started out with ordinary people discussing their ideas is now left to just a handful of specialists. Science has become too technical for the common person to understand. Scientists communicate with each other in a language that seems foreign to most of us and with mathematical formulas that seem meaningless. Yet, they believe if a unified theory were found, it would be simple enough that even the common person would be able to grasp it. The unified theory would be so compelling there would be no doubt of its accuracy.

What would happen if a common layman were the one to formulate and propose the unified theory? Would scientists even listen or be able to understand the layman's language?

I have formulated some of my insights and have developed a model that may contribute to the quest to find a unified theory. They are simple ideas which are built on simple structures that may have been part of the blueprint for the creation of the universe. I best communicate the model through visuals so it should be understood by most laymen.

Will scientists be able to understand?

Introduction

You are probably asking yourself what God's Field is. God's Field is what I define as the portion of the Higgs Field located nearest the point where the Big Bang took place. It encompasses the small area where the 118 elements along with 2 other elements I define as God's Particles were formed. In my previous book, *God's Particle and Elements: Core of the Universe. A Theory of Everything*, I presented a model of creation which showed the 120 elements were created in this small area. In this book, I incorporate God's Field into the model and show it too was created in this small area of the universe as it came into existence.

In Chapter 1, I set the stage for introducing God's Field into the model by reviewing concepts presented in my previous book. In Chapters 2 and 3, I develop innovative periodic tables by pulling the elements from their location at the Big Bang and placing them directly on the new periodic tables. In Chapter 4, I show the basic matrices that provide the structure for God's Field. In Chapter 5, I show step by step visuals as a section of God's Field shifts so two matrices can combine. This shifting or symmetry breaking transforms God's Field from a balanced field to an unbalanced one. In Chapter 6, I link the elements to God's Field and show that fundamental particles, such as quarks and electrons, are associated with each element.

Chapter 1: Elements Created at the Big Bang

How does God's Field fit into the model of creation I presented in my previous book *God's Particle and Elements*? In this chapter, I will give the basic framework of the model of creation presented in that book. In the following chapters, I will fit God's Field onto this framework.

In my previous book, I presented a theory of everything by developing a model of the creation of the universe, based on the atomic structures of the 118 known elements. While developing the model 2 additional hypothetical elements, which some may consider as God's particles, had to be added and were both assigned an atomic number of 0. The model showed how hidden universes were created as these 120 elements separated into waves of particles and anti-particles forming four sections of the universe. The same model was further developed and provided the missing link to unify the micro world of quantum mechanics with the macro world of general relativity by presenting a theory of quantum gravity.

Figure 1   Elements created at Big Bang

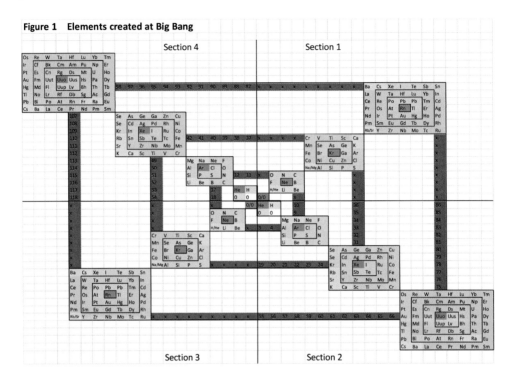

Figure 1 shows the 120 elements as they were created during the first moments of the universe. The center of the image is the point where the Big Bang took place. Families of elements radiate out from this point forming four different sections of the universe.

Figure 2   Sections of universe mirror each other

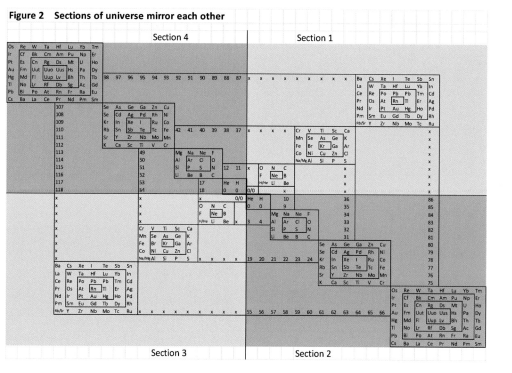

Figure 2 shows sections 1 and 3 are mirror images of each other while sections 2 and 4 mirror each other. We live in just one of the sections of the universe and the other three sections are hidden from us.

Figure 3   Families of elements form inside squares

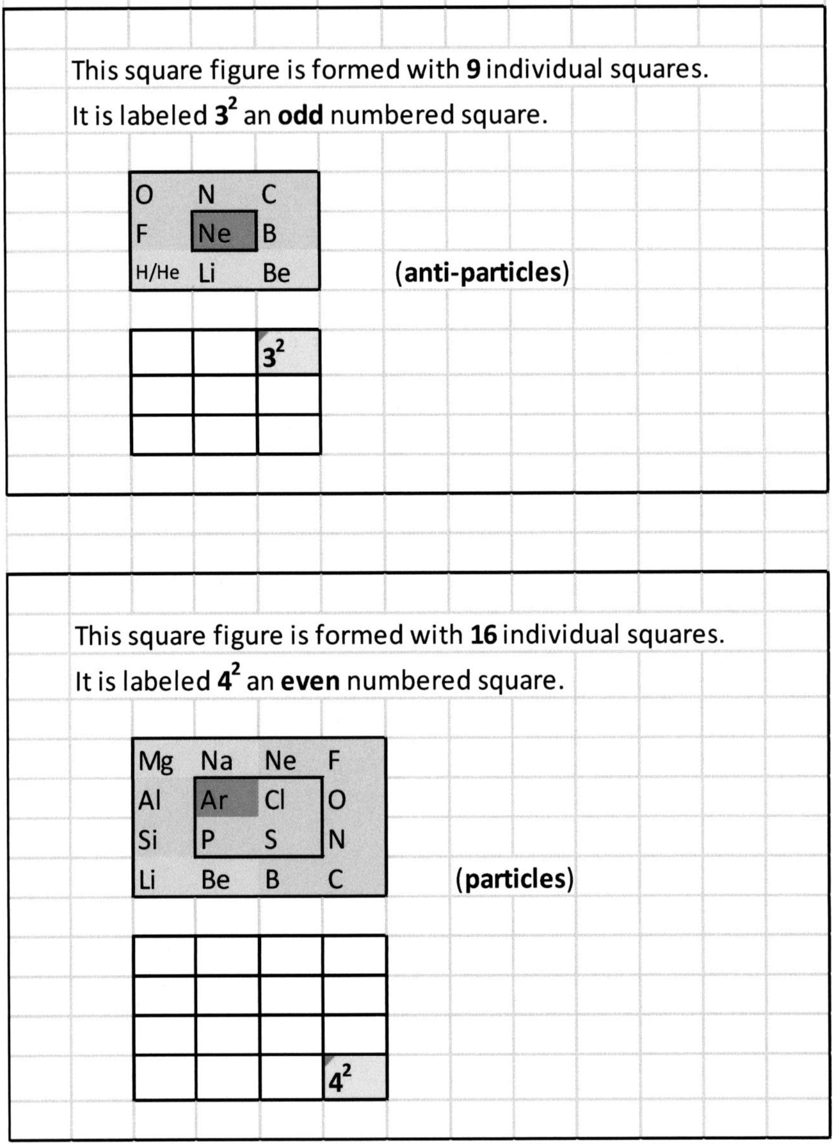

This square figure is formed with **9** individual squares.
It is labeled **3²** an **odd** numbered square.

O	N	C
F	Ne	B
H/He	Li	Be

(anti-particles)

		3²

This square figure is formed with **16** individual squares.
It is labeled **4²** an **even** numbered square.

Mg	Na	Ne	F
Al	Ar	Cl	O
Si	P	S	N
Li	Be	B	C

(particles)

			4²

Families of elements are formed inside **odd** and **even** squares as shown in Figure 3.

Figure 4   Even and odd squares fill different sections of universe

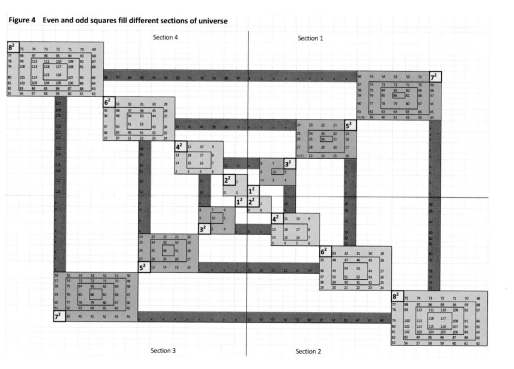

Figure 4 shows sections 1 and 3 of the universe, which mirror each other, are formed with families of elements located inside **odd** numbered squares. I define the elements located inside squares labeled 1^2, 3^2, 5^2, and 7^2 as **anti-particles**. Sections 2 and 4 of the universe, which mirror each other, are formed with families of elements inside **even** numbered squares. The group of elements inside the squares labeled 2^2, 4^2, 6^2, and 8^2 are defined as **particles**. If particles and their anti-particles came into contact with each other, they would annihilate each other. Being separated into different sections of the universe prevents this from happening.

Chapter 2: Periodic Table (Even Squares)

Now imagine designing a periodic table directly from a model of the creation of the universe. Such a periodic table would have a deep meaning and important significance because it would be linked to such an incredible event. In contrast, the periodic table we currently use today only provides us with basic information. Today's periodic table shows the elements' atomic number, mass, symbol, etc., but is not linked to anything else of greater importance.

In this chapter, I will design a periodic table using the elements in the **even** numbered square families located in sections 2 and 4 of the universe. I will take the position of each family directly from the model of creation and place it on a table. I will then unbundle the elements into rows and columns, thus presenting a new periodic table linked to the Big Bang from the perspective of the **even** numbered square families.

.

Figure 5   Even squares in sections 2 and 4

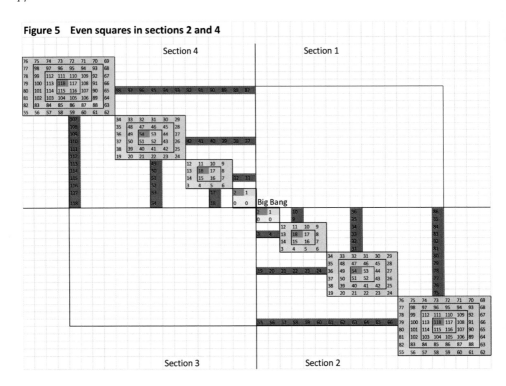

Figure 5 shows the atomic number of the elements which are located inside the **even** numbered squares and their position relative to the Big Bang in the center of figure.   The 2 elements with atomic number 0 (God's particles) are located closest to the area of the Big Bang. These 2 elements, along with the other 118 elements located in the **even** numbered squares, extend out into sections 2 and 4 of the universe.

Figure 6   Even squares in section 2

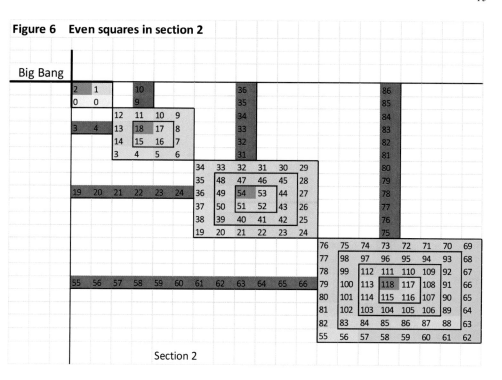

Since we live in just one section of the universe, I will use just one section of it in designing the periodic table. Figure 6 shows the position of the elements in section 2 of the universe relative to the Big Bang. The 120 elements are located inside four different **even** squares forming four families of elements. Notice each of the four families has a stable noble gas in the center.

The smallest family of elements located closest to the big bang is made of elements with atomic numbers 0-2.

12	11	10	9
13	18	17	8
14	15	16	7
3	4	5	6

The next family is formed with elements having atomic numbers 3-18.

34	33	32	31	30	29
35	48	47	46	45	28
36	49	54	53	44	27
37	50	51	52	43	26
38	39	40	41	42	25
19	20	21	22	23	24

This is followed with the next family composed of elements having atomic numbers 19-54.

76	75	74	73	72	71	70	69
77	98	97	96	95	94	93	68
78	99	112	111	110	109	92	67
79	100	113	118	117	108	91	66
80	101	114	115	116	107	90	65
81	102	103	104	105	106	89	64
82	83	84	85	86	87	88	63
55	56	57	58	59	60	61	62

The largest family located the farthest from the Big Bang is made with elements having atomic numbers 55-118.

Figure 7   Section 2 placed on a table

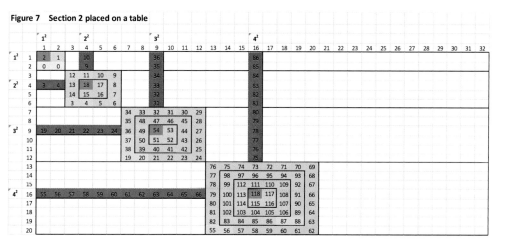

By placing section 2 of the universe on a table which has 20 rows and 32 columns I obtain the image shown in Figure 7. Notice, the noble gases located in the center of each square, line up perfectly with a square value both at the top and far left of the image. For example, the 3rd square showing noble gas Xenon, with atomic number 54, lines up perfectly with 3^2 both at the top and far left of the image. This perfect alignment of all noble gases with their square values shows the symmetry of my model and may be seen as an indirect way to validate it.

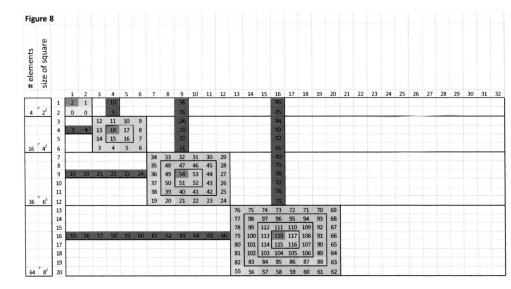

Figure 8

The next step in designing the new periodic table is to add two columns on the left of the table. As seen in Figure 8, the 1st column shows the total number of elements in each of the four families of elements. The 2nd column shows the size of each square, which is also the equivalent of the square value of the number of elements in each square. For example, the 3rd square showing elements with atomic numbers 19-54 has a total of 36 elements which is equivalent to 6^2.

Figure 9   Elements are unbundled on table

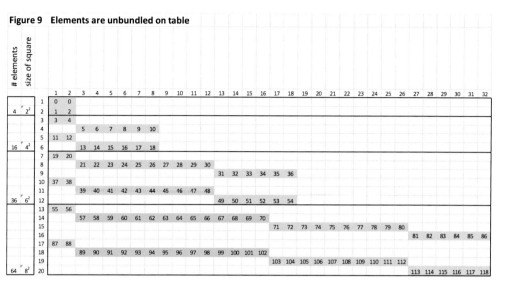

Next, I need to unbundle the elements in the four families and place them in rows on the periodic table. This is shown in Figure 9. The elements are arranged by their atomic number. Each row has 2, 6, 10, or 14 elements in it, which corresponds to the maximum number of elements allowed in each s, p, d, and, f orbital.

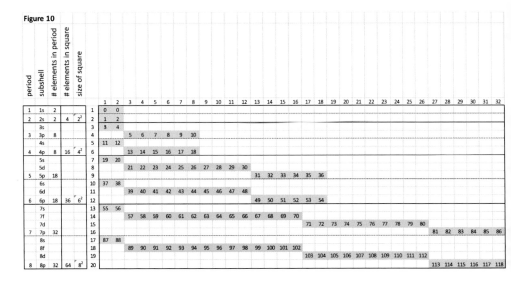

I now add three additional columns on the far left of the table as shown in Figure 10. The first column shows the different periods in the periodic table, the second column shows the subshells of the elements, and the third column shows the number of elements in each period.

The four families of elements of the **even** numbered squares are located in sets of periods ending with an **even** number. The families are located in periods 1-2, 3-4, 5-6, and 7-8.

It is important to note the sequence of numbers in the third column. The sequence is 2, 2, 8, 8, 18, 18, 32, and 32; which is the same sequence that exists in the periodic law of chemistry. The periodic law of chemistry states that chemical and physical properties of elements tend to occur in a systematic way with increasing atomic number following the sequence of 2, 8, 8, 18, 18, 32, and 32.

Once again, my model is indirectly validated because it respects the periodic law of chemistry. It is able to respect the periodic law of chemistry because the new atomic model I am presenting does not have overlapping energy levels as opposed to the current atomic model. The newly defined periods and subshells for the elements, shown in Figure 10, are a natural consequence of the proposed model of creation being presented.

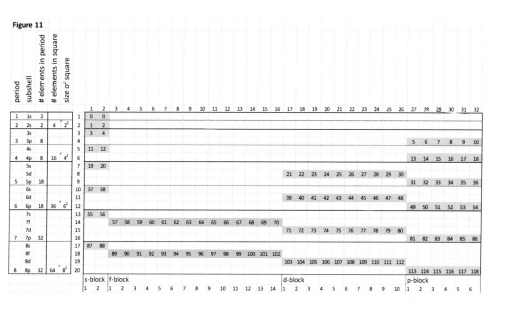

Figure 11

Elements can also be arranged in a periodic table according to their corresponding s, f, d, and p blocks. Figure 11 shows this arrangement. By looking at the bottom of the periodic table shown above, you will see the number of elements that make up each of the different blocks. The s-block has 2 elements, f-block has 14 elements, d-block has 10 elements, and the p-block has 6 elements.

Figure 12   Periodic table designed from model of creation

period	1	2	3	4	5	6	7	8	9	10	11	12	13	14	15	16	17	18	19	20	21	22	23	24	25	26	27	28	29	30	31	32
1	G1 0	G2 0																														
2	H 1	He 2																														
3	Li 3	Be 4																									B 5	C 6	N 7	O 8	F 9	Ne 10
4	Na 11	Mg 12																									Al 13	Si 14	P 15	S 16	Cl 17	Ar 18
5	K 19	Ca 20															Sc 21	Ti 22	V 23	Cr 24	Mn 25	Fe 26	Co 27	Ni 28	Cu 29	Zn 30	Ga 31	Ge 32	As 33	Se 34	Br 35	Kr 36
6	Rb 37	Sr 38															Y 39	Zr 40	Nb 41	Mo 42	Tc 43	Ru 44	Rh 45	Pd 46	Ag 47	Cd 48	In 49	Sn 50	Sb 51	Te 52	I 53	Xe 54
7	Cs 55	Ba 56	La 57	Ce 58	Pr 59	Nd 60	Pm 61	Sm 62	Eu 63	Gd 64	Tb 65	Dy 66	Ho 67	Er 68	Tm 69	Yb 70	Lu 71	Hf 72	Ta 73	W 74	Re 75	Os 76	Ir 77	Pt 78	Au 79	Hg 80	Tl 81	Pb 82	Bi 83	Po 84	At 85	Rn 86
8	Fr 87	Ra 88	Ac 89	Th 90	Pa 91	U 92	Np 93	Pu 94	Am 95	Cm 96	Bk 97	Cf 98	Es 99	Fm 100	Md 101	No 102	Lr 103	Rf 104	Db 105	Sg 106	Bh 107	Hs 108	Mt 109	Ds 110	Rg 111	Cn 112	Uut 113	Fl 114	Uup 115	Lv 116	Uus 117	Uuo 118

s-block		f-block														d-block										p-block					
1	2	1	2	3	4	5	6	7	8	9	10	11	12	13	14	1	2	3	4	5	6	7	8	9	10	1	2	3	4	5	6

The final step in designing the new period table from the model of creation is to add the symbols for each element. I assigned the symbols G1 and G2 to the 2 elements with atomic number 0 (symbol G for God). Figure 12 shows the new periodic table with the 120 elements distributed into 8 different periods.

Chapter 3: Periodic Table (Odd Squares)

In this chapter, I will design a periodic table using the elements in the **odd** numbered square families located in sections 1 and 3 of the universe. I will follow exactly the same steps that were used in the previous chapter. I will take the position of each family directly from the model of creation and place it on a table. I will then unbundle the elements into rows and columns, thus presenting a new periodic table linked to the Big Bang from the perspective of the **odd** numbered square families.

Figure13   Odd squares in sections 1 and 3

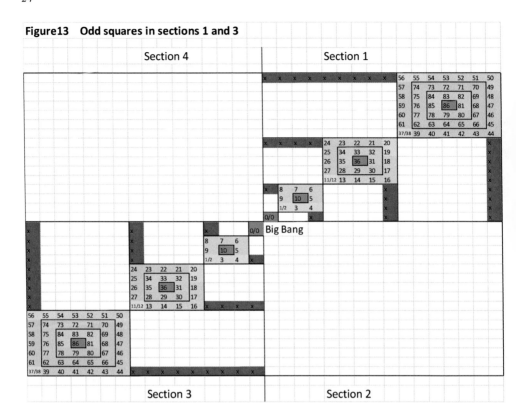

Figure 13 shows the atomic number of the elements which are located inside the **odd** numbered squares. The 2 elements with atomic number 0 (God's particles) are located at the center.   These 2 elements, along with the other 86 elements located in the **odd** numbered squares, extend out into sections 1 and 3 of the universe.

Figure 14   Odd squares in section 2

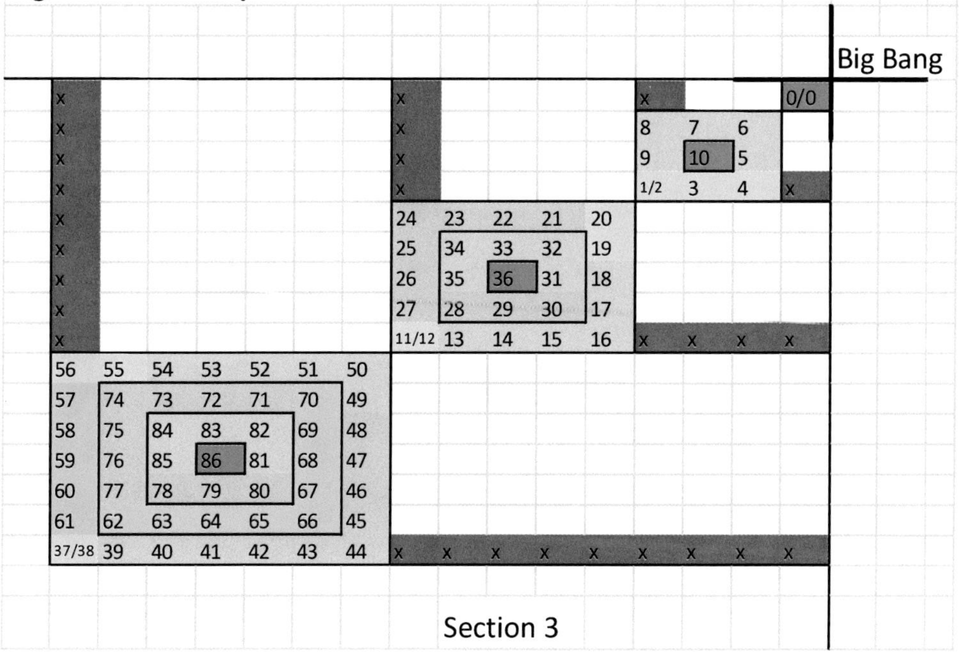

Big Bang

				0/0
8	7	6		
9	10	5		
1/2	3	4		

24	23	22	21	20
25	34	33	32	19
26	35	36	31	18
27	28	29	30	17
11/12	13	14	15	16

56	55	54	53	52	51	50
57	74	73	72	71	70	49
58	75	84	83	82	69	48
59	76	85	86	81	68	47
60	77	78	79	80	67	46
61	62	63	64	65	66	45
37/38	39	40	41	42	43	44

Section 3

Since we live in just one section of the universe, I will use just one section in designing the periodic table. Figure 14 shows the position of the elements in section 3 of the universe relative to the Big Bang. The 86 elements are located inside four different **odd** squares forming four families of elements. Notice each of the four families has a stable noble gas in the center.

0/0

The smallest family of elements located closest to the big bang is made with elements having an atomic number of 0.

8	7	6
9	10	5
1/2	3	4

The next family is formed with elements having atomic numbers 1-10.

24	23	22	21	20
25	34	33	32	19
26	35	36	31	18
27	28	29	30	17
11/12	13	14	15	16

This is followed with the next family composed of elements having atomic numbers 11-36.

56	55	54	53	52	51	50
57	74	73	72	71	70	49
58	75	84	83	82	69	48
59	76	85	86	81	68	47
60	77	78	79	80	67	46
61	62	63	64	65	66	45
37/38	39	40	41	42	43	44

The largest family located the farthest from the big bang is made with elements having atomic numbers 37-86.

Figure 15 Section 3 placed on a table

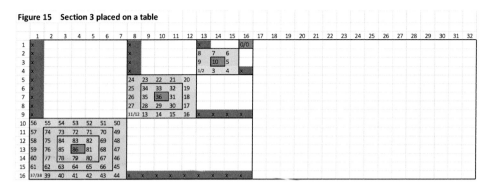

By placing section 3 of the universe on a table which has 16 rows and 32 columns I obtain the image shown in Figure 15.

Figure 16

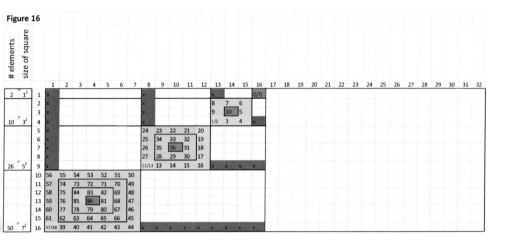

The next step in designing the new periodic table is to add two columns on the left of the table. As seen in Figure 16, the 1st column shows the total number of elements in each square. The 2nd column shows the size of each of these squares. For example, the square with atomic numbers 11-36 has 26 elements in it and a size of 5^2.

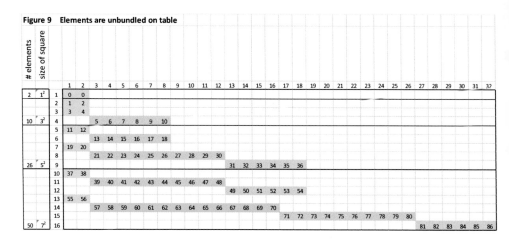

Next, I need to unbundle the elements in the four families and place them in rows on the periodic table. This is shown in Figure 9. The elements are arranged by their atomic number. Each row has 2, 6, 10, or 14 elements in it, which corresponds to the maximum number of elements allowed in each s, p, d, and, f orbital.

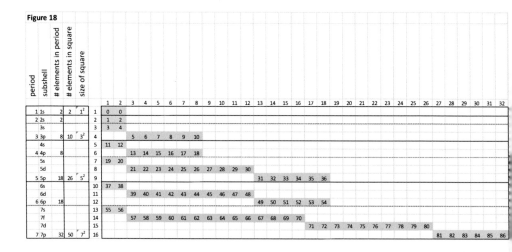

I now add three additional columns on the far left of the table as shown in Figure 18. The first column shows the different periods in the periodic table, the second column shows the subshells of the elements, and the third column shows the number of elements in each period.

The four families of elements of the **odd** numbered squares are located in sets of periods ending with an **odd** number. The families are located in periods 1, 2-3, 4-5, and 6-7.

It is important to note the sequence of numbers in the second column. The sequence is 2, 2, 8, 8, 18, 18, and 32; which is the same sequence that exists in the periodic law of chemistry. The periodic law of chemistry states that chemical and physical properties of elements tend to occur in a systematic way with increasing atomic number following the sequence of 2, 8, 8, 18, 18, and 32.

Once again, my model is indirectly validated because it respects the periodic law of chemistry. It is able to respect the periodic law of chemistry because the new atomic model I am presenting does not have overlapping energy levels as opposed to the current atomic model. The newly defined periods and subshells for the elements, shown in Figure 18, are a natural consequence of the proposed model of creation being presented.

Figure 19

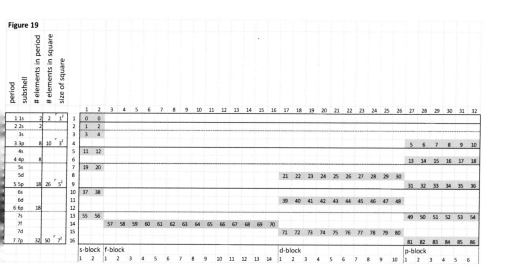

Elements can also be arranged in a periodic table according to their corresponding s, f, d, and p blocks. Figure 19 shows this arrangement. By looking at the bottom of the periodic table shown above, you will see the number of elements that make up each of the different blocks, which form the periodic table. The s-block has 2 elements, f-block has 14 elements, d-block has 10 elements, and the p-block has 6 elements.

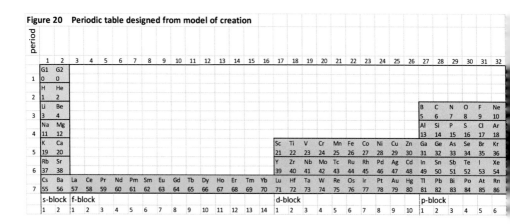

Figure 20   Periodic table designed from model of creation

The final step in designing the new period table from the model of creation is to add the symbols for each element. I assigned the symbols G1 and G2 to the 2 elements with atomic number 0 (symbol G for God). Figure 20 shows the new periodic table with the 88 elements distributed into 7 different periods.

Chapter 4: God's Field and Matrices

Many scientists believe the Higgs Field wasn't created by a process. They believe it is something that is just "there", the same way the electric field of nature is just there, always and everywhere. I completely disagree with that view. In this chapter, I will show God's Field did originate through a process. The creation of the universe followed a certain process or steps. The steps can be unveiled through logic using simple mathematical models and structures.

God's Field has an underlying structure. The underlying structure is made with simple matrices. In this chapter, I will show you how I "unveiled" the underlying structure the universe is built on.

I started by using simple sets of matrices. A matrix is a collection of numbers or data which are arranged into rows and columns. For example, a matrix with 2 rows and 4 columns is called a 2x4 matrix. A simple 2x4 matrix can be made by flipping a coin and recording the results needed to get 2 consecutive heads. On average, how many times would you have to flip a coin to get 2 consecutive heads? The answer is 8 times.

Figure 21				
2x4 matrix				
	1	2	3	4
1	heads	heads	tails	tails
2	heads	tails	heads	tails

Figure 21 shows the 4 possible results of flipping a coin 8 times in order to get 2 consecutive heads. The 4 possible outcomes are shown in the 4 columns: heads/heads, heads/tails, tails/heads, and tails/tails.

Figure 22				
2x4 matrix				
	1	2	3	4
1	heads	heads	tails	tails
2	heads	tails	heads	tails
	1	2	3	4
1	+	+	-	-
2	+	-	+	-

The symbol "+" can be substituted for "heads" and the symbol "-" can be substituted for "tails" on the 2x4 matrix as shown in Figure 22.

Figure 23   Basic matrices used to build universe

1x1 matrix

	1
1	+

1x2 matrix

	1	2
1	+	-

2x4 matrix

	1	2	3	4
1	+	+	-	-
2	+	-	+	-

3x8 matrix

	1	2	3	4	5	6	7	8
1	+	+	+	+	-	-	-	-
2	+	+	-	-	+	+	-	-
3	+	-	+	-	+	-	+	-

4x16 matrix

	1	2	3	4	5	6	7	8	9	10	11	12	13	14	15	16
1	+	+	+	+	+	+	+	+	-	-	-	-	-	-	-	-
2	+	+	+	+	-	-	-	-	+	+	+	+	-	-	-	-
3	+	+	-	-	+	+	-	-	+	+	-	-	+	+	-	-
4	+	-	+	-	+	-	+	-	+	-	+	-	+	-	+	-

Figure 23 shows the five matrices that I used to unveil God's Field. They are the simple fundamental structures that lie at the core of the universe. Imagine building the whole universe starting out with just a simple 1x1 matrix! However, if you were to build a house, you too would start with just one brick and build on that. Notice, the length of each matrix (brick) shown above is double the length of the preceding one. The lengths of 1, 2, 4, 8, and 16 form a **balanced** set of matrices.

As you recall the universe is divided into four different sections. In order for the matrices to cover all the sections of the universe, they have to be increased by a factor of four.

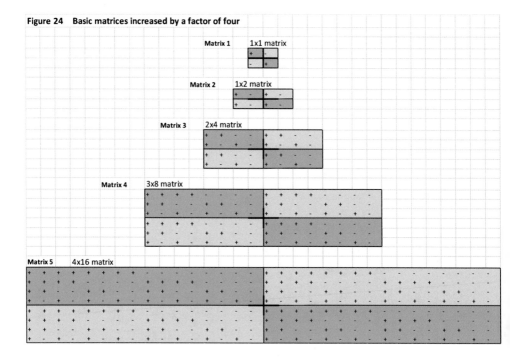

Figure 24   Basic matrices increased by a factor of four

Figure 24 shows the **5** matrices after being increased by a factor of four. So now, we have larger matrices which will cover a larger area. As you can see, the larger matrices are made by combining four of the smaller matrices. Notice the 1x1 matrix is still the smallest unit even though it is now located inside a larger one.

Let's return for a moment to the analogy of using bricks for building a house. Where do the bricks come from? They are made with a mixture of raw materials including clay, shale, and water. This mixture is poured into a mold before firing it in a kiln. The mold provides the basic structure for the shape of a brick.

God's Field is like an invisible mold. It is present everywhere and provides the basic structure when energy is converted into mass. The smallest known particles formed in this mold during this conversion of energy to mass are called fundamental particles.

Which elements or fundamental particles are located inside the smallest 1x1 matrix of the model of creation?

In Chapter 3, I showed the smallest family of elements located closest to the Big Bang is made with elements having an atomic number of 0. I assigned the symbols G1 and G2 to the 2 elements with atomic number 0 (symbol G for God).

What is the whole universe built on? That's right; the "brick" with God's particles in it, formed in the smallest mold of God's Field.

Let's return to the image shown in Figure 24. This set of matrices gives us the first glimpse of God's Field as a **balanced** field. If you think of the center of each of the **5** individual matrix as the point where the Big Bang took place, you can see there is a field, shown with symbols of "+" and "-", extending out from this point to each of the four sections of the universe.

In Chapter 1, I showed an image of the elements also spreading out into the four sections of the universe. God's Field and the elements both extend out into four sections of the universe because both were created at the Big Bang. The mold provided by God's Field was the invisible structure present when energy converted to mass forming the fundamental particles after the Big Bang.

Figure 25   Elements spread out into 4 sections of the universe

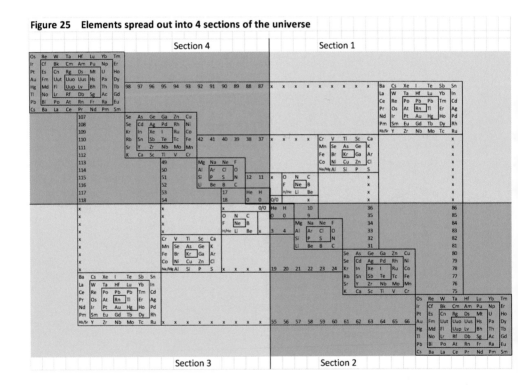

Figure 25 is the image shown in Chapter 1. It shows the **4** families of elements in each of the odd and even sections of the universe as they extend out into the universe. I will end this chapter with a chart showing the atomic numbers of these **4** families of elements.

Figure 26   Families of elements

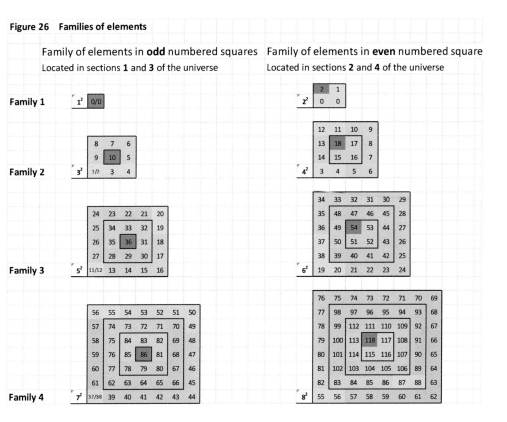

Figure 26 shows the atomic numbers of the elements in the **4** families forming the odd numbered squares and the **4** families forming the even squares. Each family has a stable noble gas in the center.

Chapter 5: God's Field Shifts

In the previous chapter, I was able to establish a link between God's Field and the **4** family of elements. I did this by showing visually they both originated at the Big Bang and extended out into the four sections of the universe. I also drew attention to the fact that the number of matrices in a balanced field did not coincide with the number of families of elements. There are **5** field matrices in a balanced field and just **4** families of elements.

In order to design a symmetrical model of creation, I need to make those numbers match. In this chapter, I will share with you the process I followed in order make the numbers match. As you will see, achieving a goal of having **4** field matrices and **4** families of elements was not an easy task, yet the universe has secrets just waiting to be unveiled.

I decided I would begin by trying to eliminate one of the **5** field matrices. I decided to try combining two field matrices into one. By combining two matrices, instead of having **5** field matrices, I would end up with the desired **4** field matrices. The question now was which two matrices to combine.

In order to choose the two matrices to combine, I decided to look at number patterns in my model for clues that might help me with the decision.

Figure 27	Number patterns	
Matrix 1	**1**x1 matrix	
Matrix 2	**1**x2 matrix	*numeral 1 is repeated*
Matrix 3	**2**x4 matrix	
Matrix 4	**3**x8 matrix	
Matrix 5	**4**x16 matrix	

I looked at the number patterns of the original small matrices (bricks) which were used to form larger matrices (Figure 23). As shown in Figure 27, the sequence of rows in the matrices is 1, **1**, 2, 3, and 4. The numeral 1 is repeated in Matrix 2. Since this repetition broke a pattern, I chose Matrix 2 to be one of the matrices I would use to combine with another matrix. I now needed to figure out the other matrix to use.

I decided to get another view of the situation by placing all 5 matrices on a table.

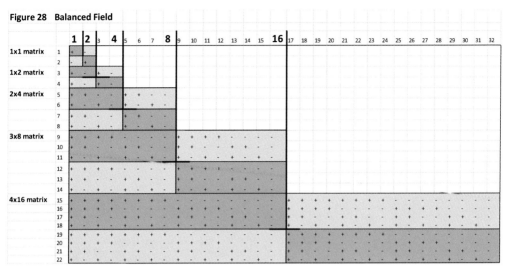

Figure 28   Balanced Field

I placed the 5 matrices on a table with 22 rows and 32 columns as shown in Figure 28. I saw this placement created a **balanced field,** since the centers of the matrices are located in the 1st, 2nd, 4th, 8th, and 16th columns. Each center is double the distance from the previous center. Next, I decided to remove the 1x2 matrix (Matrix 2) from the table since I had already chosen it as a candidate to combine with another matrix.

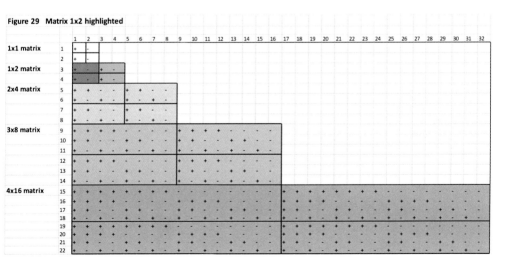

Figure 29   Matrix 1x2 highlighted

Figure 29 shows the 1x2 matrix highlighted before I removed it from the table.

Figure 30   Matrix 1x2 temporarily removed

		1	2	3	4	5	6	7	8	9	10	11	12	13	14	15	16	17	18	19	20	21	22	23	24	25	26	27	28	29	30	31	32
1x1 matrix	1	+	-																														
	2	+	-																														
	3																																
	4																																
2x4 matrix	5	+	+	-	-	+	+	-	-																								
	6	+	-	+	-	+	-	+	-																								
	7	+	+	-	-	+	+	-	-																								
	8	+	-	+	-	+	-	+	-																								
3x8 matrix	9	+	+	+	+	-	-	-	-	+	+	+	+	-	-	-	-																
	10	+	+	-	-	+	+	-	-	+	+	-	-	+	+	-	-																
	11	+	-	+	-	+	-	+	-	+	-	+	-	+	-	+	-																
	12	+	+	+	+	-	-	-	-	+	+	+	+	-	-	-	-																
	13	+	+	-	-	+	+	-	-	+	+	-	-	+	+	-	-																
	14	+	-	+	-	+	-	+	-	+	-	+	-	+	-	+	-																
4x16 matrix	15	+	+	+	+	+	+	+	+	-	-	-	-	-	-	-	-	+	+	+	+	+	+	+	+	-	-	-	-	-	-	-	-
	16	+	+	+	+	-	-	-	-	+	+	+	+	-	-	-	-	+	+	+	+	-	-	-	-	+	+	+	+	-	-	-	-
	17	+	+	-	-	+	+	-	-	+	+	-	-	+	+	-	-	+	+	-	-	+	+	-	-	+	+	-	-	+	+	-	-
	18	+	-	+	-	+	-	+	-	+	-	+	-	+	-	+	-	+	-	+	-	+	-	+	-	+	-	+	-	+	-	+	-
	19	+	+	+	+	+	+	+	+	-	-	-	-	-	-	-	-	+	+	+	+	+	+	+	+	-	-	-	-	-	-	-	-
	20	+	+	+	+	-	-	-	-	+	+	+	+	-	-	-	-	+	+	+	+	-	-	-	-	+	+	+	+	-	-	-	-
	21	+	+	-	-	+	+	-	-	+	+	-	-	+	+	-	-	+	+	-	-	+	+	-	-	+	+	-	-	+	+	-	-
	22	+	-	+	-	+	-	+	-	+	-	+	-	+	-	+	-	+	-	+	-	+	-	+	-	+	-	+	-	+	-	+	-
1x2 matrix		+	-	+	-																												
		+	-	+	-																												

Figure 30 shows the 1x2 matrix after I removed it from the table and temporarily placed it at the bottom of the table, until I would hopefully be able to combine it with another matrix.

Figure 31 Centers of matrices

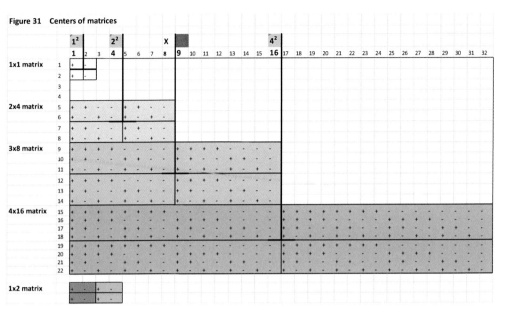

I took a close look at the centers of the four remaining matrices. In Figure 31, you can see they are located in columns 1, 4, **8**, and 16. The top row shows the sequence of 1^2, 2^2, **X**, and 4^2 for the centers. The center of the 3x8 matrix is off. The center should be located in column 9 (3^2) so the sequential values of 1^2, 2^2, 3^2, and 4^2 can be written in the top row. A section of the 3x8 matrix needs to be shifted to the right to form a new center located in column 9 (3^2).

I now had found the other matrix I would use for combining two matrices. I would combine the 1x2 matrix with the off centered 3x8 matrix. By combining these two matrices, I would end up with a new 3x9 matrix.

I now needed to shift a section of the 3x8 matrix to the right.

Figure 32   Section of 3x8 matrix selected

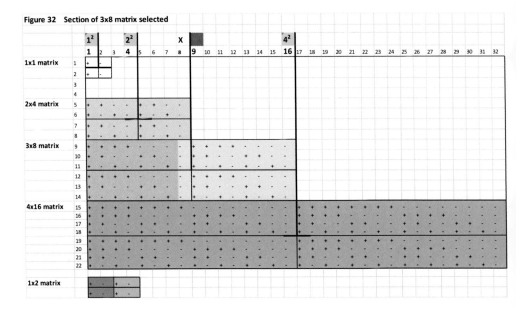

Figure 32 shows the section of the 3x8 matrix that I decided to shift to the right.

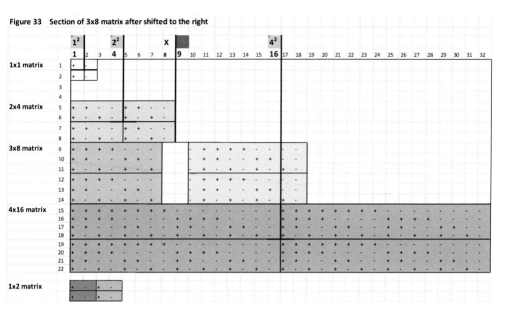

Figure 33 Section of 3x8 matrix after shifted to the right

Figure 33 shows the new location of the 3x8 matrix section after I shifted it to the right.

Figure 34   1x2 matrix inserted into 3x8 matrix

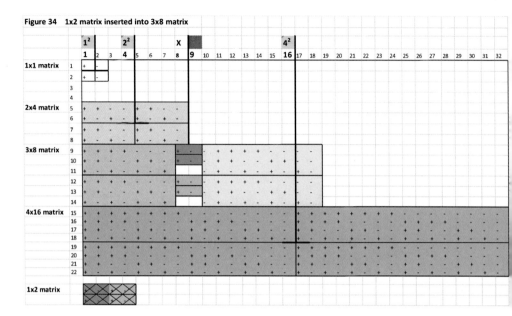

I then inserted the 1x2 matrix into the empty space that was created by shifting the 3x8 matrix to the right. Figure 34 shows this. Thus, I successfully had combined two matrices, which was the task I had set out to do.

Figure 35   Centers of matrices

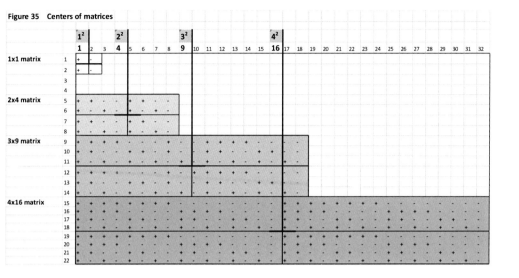

By looking at Figure 35, you will see the center of the newly formed 3x9 matrix (replacing the former 3x8 matrix) is now located at 9 (3^2).   The sequence of the centers of the four matrices is now 1^2, 2^2, 3^2, and 4^2. Perfect.

Figure 36   God's Field forms an unbalanced field

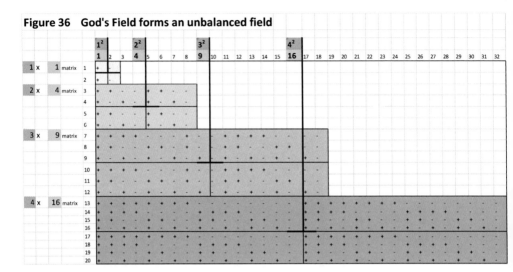

God's Field is finally unveiled by eliminating the two rows on the table where the 1x2 matrix was originally located. Figure 36 shows the completed God's Field on a table with 20 rows and 32 columns. The centers of the **4** matrices are now located in the 1st, 4th, 9th and 16th column forming an **unbalanced field.** Each center (1, 4, 9, and 16) is no longer double the distance from the preceding one.

The term **unbalanced** is usually perceived to be a negative attribute. However, in this case, it is actually extremely positive because the new centers are now located at values with 1^2, 2^2, 3^2, and 4^2. These squared values provide a perfect structure for the **4** families of elements to fit on, since all the families of elements are located inside shapes with squared values.

Figure 37 Families of elements fit PERFECTLY on God's Field

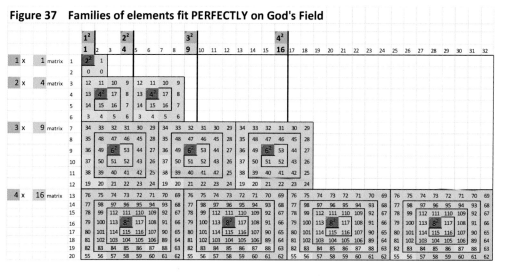

Figure 37 shows the splendor of the perfect match of families of elements with their corresponding field matrices.

It is truly mathematically amazing the shift that took place with regards to the center of each matrix to allow this perfect fit to take place. Going from a **balanced field** with numbers at the centers of 1, 2, 4, 8, and 16 (each number doubling the previous) to an **unbalanced field** with 1^2, 2^2, 3^2, and 4^2 (series of consecutive squares) at the center is astounding to say the least.

The perfect match shown in Figure 37 of the families of elements with their corresponding field matrices is another way in which the model of creation, I have been presenting, is indirectly validated. It seems to be so compelling there is little doubt of its accuracy.

Chapter 6: Fundamental Particles Emerge from God's Field

In the previous chapter, I reduced the original **5** field matrices to **4** field matrices and placed them on a table with 20 rows and 32 columns. I labeled the table; God's Field forms an unbalanced field (Figure 36).

I now need to combine that table with the periodic table designed in Chapter 2 (Figure 10). Remember the periodic table was designed by directly using section 2 of the universe from the image of the Big Bang, so it too, is linked to the moment of creation.

Figure 36   God's Field forms an unbalanced field

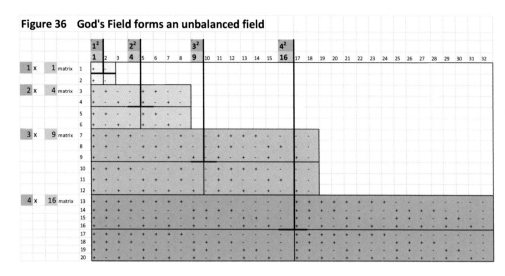

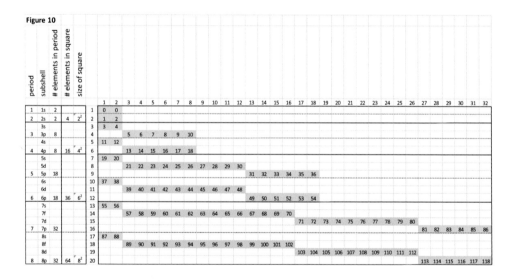

The tables in Figure 36 and Figure 10 are the two tables I will combine. Since they have the same number of rows and columns, the data from the two tables can be combined forming a new table

Figure 38   God's Field combined with elements on periodic table

Matrix	Row	1²		2²						3²							4²															
		1	2	3	4	5	6	7	8	9	10	11	12	13	14	15	16	17	18	19	20	21	22	23	24	25	26	27	28	29	30	31	32
1x1 matrix	1	0	0																														
	2	1	2																														
2x4 matrix	3	3	4	-	-	+	+	-	-																								
	4	+	-	5	6	7	8	9	10																								
	5	11	12	-		+	+	-	-																								
	6	+	-	13	14	15	16	17	18																								
3x9 matrix	7	19	20	+	+	-	-	-	-	+	-	+	+	+	+	-	-	-	-														
	8	+	+	21	22	23	24	25	26	27	28	29	30	-	-	+	+	-	-														
	9	+	-	+	-	+	-	+		-	+	-		31	32	33	34	35	36														
	10	37	38	+	+	-	-	-	+	-	-	+	+	+	+	-	-	-	-														
	11	+	+	39	40	41	42	43	44	45	46	47	48	-	-	+	+	-	-														
	12	+	-	+	-	+	-	+		-	+	-		49	50	51	52	53	54														
4x16 matrix	13	55	56	+	+	+	+	+	+	-	-	-	-	-	-	+	+	+	+	+	+	+	+	+	+	-	-	-	-	-	-	-	-
	14	+	+	57	58	59	60	61	62	63	64	65	66	67	68	69	70	+	+	+	+	-	-	-	-	+	+	-	-	-	-	-	-
	15	+	+	+	+	+	+	-	-	+	+	+	+	-	-	+	+	71	72	73	74	75	76	77	78	79	80	-	-	+	+	-	-
	16	+	-	+	-	+	-	+	-	+	-	+	-	+	-	+	-	+	-	+	-	+	-	+	-	+	-	81	82	83	84	85	86
	17	87	88	+	+	+	+	+	+	-	-	-	-	-	-	+	+	+	+	+	+	+	+	+	+	-	-	-	-	-	-	-	-
	18	+	+	89	90	91	92	93	94	95	96	97	98	99	100	101	102	+	+	+	+	-	-	-	-	+	+	-	-	-	-	-	-
	19	+	+	+	+	+	+	-	-	+	+	+	+	-	-	+	+	103	104	105	106	107	108	109	110	111	112	-	-	+	+	-	-
	20	+	-	+	-	+	-	+	-	+	-	+	-	+	-	+	-	+	-	+	-	+	-	+	-	+	-	113	114	115	116	117	118

Figure 38 shows the 120 elements placed on God's Field.   The column on the left shows the **4** field matrices numbered in a sequential order of: 1x1, **2x4**, **3x9**, and **4x16**.   The number of field matrices matches the **4** families of elements found on the periodic table. The center of each matrix (shown in first row at top of image) is also numbered in a sequential order of: 1^2, 2^2, 3^2, and 4^2.   Perfect.

It is important to notice each of the 120 elements on the table is associated with its own "+" and "-" symbols on God's Field.

Figure 39   Each element is associated with its own fundamental particles

0	0
+	-

element
fundamental particles

1	2
+	-

element
fundamental particles

3	4	5	6	7	8	9	10
+	+	-	-	+	+	-	-
+	-	+	-	+	-	+	-

element
fundamental particles
fundamental particles

11	12	13	14	15	16	17	18
+	+	-	-	+	+	-	-
+	-	+	-	+	-	+	-

element
fundamental particles
fundamental particles

19	20	21	22	23	24	25	26	27	28	29	30	31	32	33	34	35	36
+	+	+	+			-	-	+	-	-	+	+	+	+	-	-	-
+	+	-	-	+	+	-	-	+	+	-	-	+	+	-	-	+	-
+	-	+	-	+	-	+	-	+	-	+	-	+	-	+	-	+	-

element
fundamental particles
fundamental particles
fundamental particles

37	38	39	40	41	42	43	44	45	46	47	48	49	50	51	52	53	54
+	+	+	+			-	-	+	-	-	+	+	+	+	-	-	-
+	+	-	-	+	+	-	-	+	+	-	-	+	+	-	-	+	-
+	-	+	-	+	-	+	-	+	-	+	-	+	-	+	-	+	-

element
fundamental particles
fundamental particles
fundamental particles

55	56	57	58	59	60	61	62	63	64	65	66	67	68	69	70	71	72	73	74	75	76	77	78	79	80	81	82	83	84	85	86
+	+	+	+	+	+	+	+	-	-	-	-	-	-	-	-	+	+	+	+	+	+	+	+	-	-	-	-	-	-	-	-
+	+	+	+			-	-	+	+			-	-	+	+	+	+	-	-	+	+	-	-	+	+	-	-	+	-		
+	+	-	-	+	+	-	-	+	+	-	-	+	+	-	-	+	+	-	-	+	+	-	-	+	+	-	-	+	-		
+	-	+	-	+	-	+	-	+	-	+	-	+	-	+	-	+	-	+	-	+	-	+	-	+	-	+	-	+	-		

element
fundamental particles
fundamental particles
fundamental particles
fundamental particles

87	88	89	90	91	92	93	94	95	96	97	98	99	100	101	102	103	104	105	106	107	108	109	110	111	112	113	114	115	116	117	118
+	+	+	+	+	+	+	+	-	-	-	-	-	-	-	-	+	+	+	+	+	+	+	+	-	-	-	-	-	-	-	-
+	+	+	+			-	-	+	+			-	-	+	+	+	+	-	-	+	+	-	-	+	+	-	-	+	-		
+	+	-	-	+	+	-	-	+	+	-	-	+	+	-	-	+	+	-	-	+	+	-	-	+	+	-	-	+	-		
+	-	+	-	+	-	+	-	+	-	+	-	+	-	+	-	+	-	+	-	+	-	+	-	+	-	+	-	+	-		

element
fundamental particles
fundamental particles
fundamental particles
fundamental particles

Figure 39 shows the "+" and "-" symbols associated with each element.

Each "+" and "-" symbol could be interpreted to be a fundamental particle such as a quark or electron. A proton is made with 2 up quarks and 1 down quark. The "+" and "-" symbols could represent these two different types of quarks. This interpretation makes God's Field more tangible, since we can now visualize something as small as a quark or electron being formed inside the mold of God's Field.

Figure 40   Sequence to form a nucleus

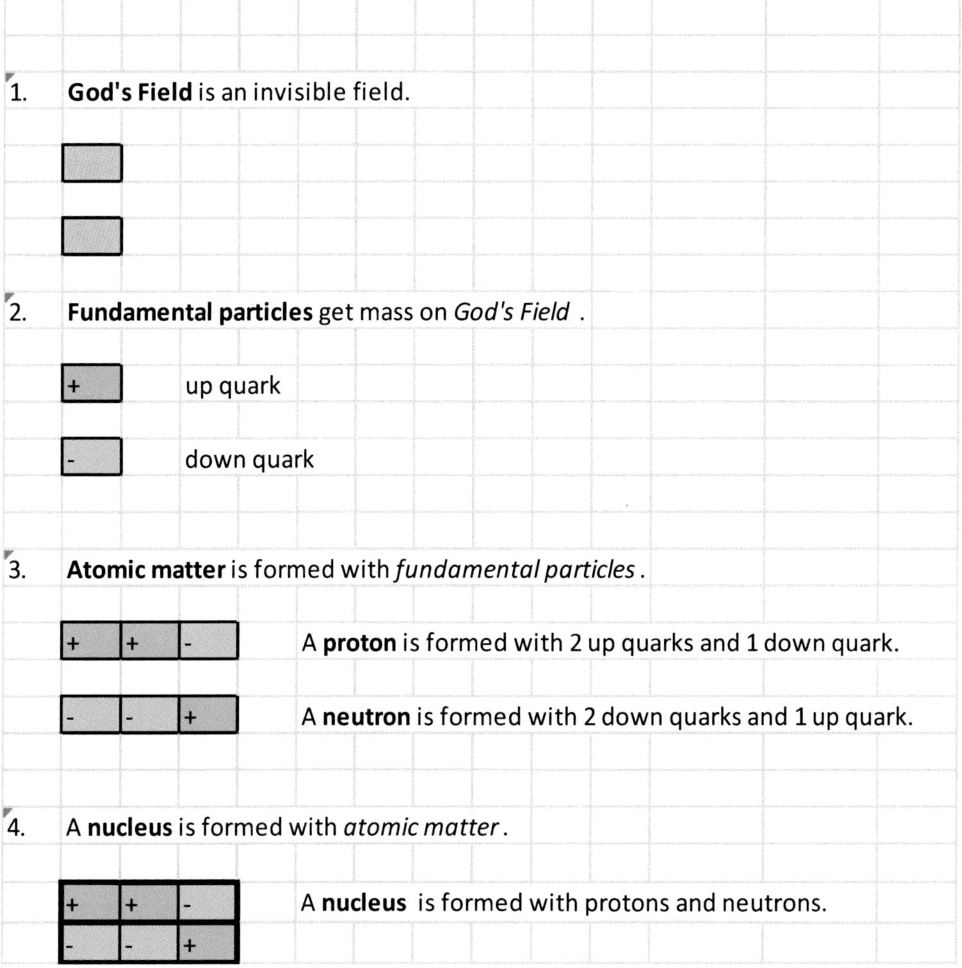

1.   **God's Field** is an invisible field.

2.   **Fundamental particles** get mass on *God's Field* .

     `+` up quark

     `-` down quark

3.   **Atomic matter** is formed with *fundamental particles* .

     `+ + -`   A **proton** is formed with 2 up quarks and 1 down quark.

     `- - +`   A **neutron** is formed with 2 down quarks and 1 up quark.

4.   A **nucleus** is formed with *atomic matter* .

     `+ + -`
     `- - +`   A **nucleus** is formed with protons and neutrons.

The sequence followed to form a nucleus of an atom can be seen in Figure 40. This sequence is congruent with current theory which theorizes the Higgs Field gives mass only to fundamental particles, such as quarks and electrons, but not to ordinary atomic matter such as protons and neutrons in a **non-zero field**.

Nobody knows for sure what happened during the earliest moments of the universe. However, I will end this book with a timeline of events as they may have happened. The timeline is based on the hypotheses of my proposed model of creation presented throughout this book and my previous one.

By assuming the **unbalanced field** formed with the **4** matrices on God's Field of my model is a **non-zero field**, I present the following timeline of events.

- A fraction of a second before the Big Bang there was a **balanced field** having a value of **zero.** (Figure 28)

- The Big Bang happened when God's Field became an **unbalanced field** with a **non-zero** value. This was when two matrices combined together. The detailed steps was shown when part of the 3x8 matrix shifted to the right and the 1x2 matrix was inserted into the empty space. (Figures 29-35)

- The **unbalanced** God's Field provided the invisible structure or mold where mass was given to fundamental particles such as quarks and electrons by the energy in the waves (ripples) that originated at the Big Bang. (Figures 38, 39)

- The fundamental particles came together to form atomic matter such as protons and neutrons. (Figure 40)

- Protons and neutrons combined to form nuclei of atoms. (Figure 40)

- Nuclei combined with electrons to form atoms of elements.

- Elements combined and formed 4 families of anti-particles in odd squares and 4 families of particles in even squares. (Figure 26)

- Families of elements spread out into four different sections of the universe orbiting the central point where the universe began. (Figures 1, 2, and 4)

About the Author

Gerald W. Scanlon was born in Saginaw, Michigan and moved to Mexico when he was 21 years old. He had a fulfilling career as a teacher and is now enjoying his golden years of retirement. He has always had a love for math and still dreams of coming up with a perfect betting system for playing blackjack.

It was actually while trying to develop a betting system that he was led to the elements on the periodic table. He saw a relationship with the way elements fill their orbitals and the patterns of wins and losses that occur while gambling. This insight led him to the amazing world of quantum mechanics where the secrets of the universe were waiting for him to unveil.

Made in the USA
Middletown, DE
17 August 2022